Winning Communication Skills for Telephone, Conference Calls

Published BY :
Abraham Hossni

About this course

Improve Communication Skills - Use the Telephone Effectively! Avoid Phone Etiquette Blunders that Annoy

Description :

Communication Skills - Telephone skills in the workplace are essential for business success. **Conference calls**, client updates, **Skype/Zoom video meetings**, and even job interviews are all conducted through the help of telephones and smart phones. In a digital era filled with endless emails and social media posts, the live human voice remains a singular power.

Are you and members of your team using the **telephone** for maximum business success?

Do you have Baby Boomers in your organization who are afraid of Skype video and Apple FaceTime and are slow to text clients and customers who may be Millennials?

Do you have Millennials on your team who don't realize that Baby Boomers expect their calls answered and their **voicemails** returned?

This course is for anyone on your team who uses a **telephone** and for everyone who wants to increase their communications and business successes with customers, clients, prospects, colleagues, investors, and bosses.

What will students achieve or be able to do after taking this course?

- Communicate effectively using the **phone** with clients, customers, colleagues and bosses
- Avoid telephone blunders that strike others as rude and unprofessional
- Speak effectively on business conference calls
- Present confidently on Skype and FaceTime calls
- **Use the telephone effectively** in all aspects of business life

Please note: this is a **telephone communications skills** course conducted by a real person who is speaking and demonstrating **communication**

skills. If you are looking for a course with lots of animation, slides, special effects, slick edits, and robotic voices, this course is not for you.

What you'll learn

- Communicate effectively using the phone with clients, customers, colleagues and bosses
- Avoid telephone blunders that strike others as rude and unprofessional
- Speak effectively on business conference calls
- Present confidently on Skype and FaceTime calls
- Use the telephone effectively in all aspects of business life

Are there any course requirements or prerequisites?

- You need a phone and voicemail

Who this course is for:

- Workers looking to build their soft skills
- Younger executives who want to communicate with everyone in their business life, including older colleagues

- Older executives who want to communicate with everyone in business, including young people

Section 1: Setting Groundwork for Telephone Communication Success

1. Quick Wins! Do THIS To Look and Sound Great on Skype Phone Calls :

Your time is valuable, so I want to give you some quick wins when it comes to how to use your phone more effectively in business.
For starters, if it's a really, really important call, perhaps you're interviewing for a job.
Perhaps you're trying to get the biggest new client for your business ever.
You want to sound your absolute best.
Here's a tip stand.
When you're doing the phone call, you'll have more breath, you'll have more energy, you'll have a greater range in your voice and try to find a mirror you can look at and actually smile.
I realize that sounds cheesy, but when you smile, it comes out in your voice.
You'll sound more upbeat, more focused, more energetic, more excited to be there.
That's what new employers want to hear.
That's also what clients and client prospects want to hear.

Another quick tip.
If you're conducting a business call through FaceTime video or Skype video, most of the time people are holding a phone down like this.
What happens is it shoots up your nose.
The brightest thing on the screen is going to be the light above you.
It's going to look horrible.
Your face will be all blackened out.
Here's a quick tip.
Hold the phone a couple of inches above your eyes.
Now, it's a much more flattering angle on your face.
The lighting will be better because you're not shooting the lights from the ceiling.
A couple of quick tips.
This course is going to be full of them, so get ready.
By the time we're done, you're going to be a real pro at using the cell phone and a regular phone in any kind of phone for every aspect of your career.

2. Never Confuse Personal Phone Use with Business Telephone Use Again:

So why is there a need for this course, people have been using phones their whole life these days,
five year olds use phones.
What's the difference for business?
Well, let me tell you, there's a huge difference between how you use a phone in business, your work life versus your personal life.
There are a lot of things that are perfectly acceptable to do with friends, family.
But if you do it in a business situation, you will lose clients, you will lose customers, you will lose promotions.
So let's clearly define what we're talking about here.
This is not how to call your best friend and say, let's go to a party.
This is about how to use the phone for business and really understanding the differences, what's appropriate,
what isn't.
If you call your best friend and you don't leave a message, you know they're going to see you call, they'll call you back.
Doesn't matter.

But if a very important prospect, your client calls you and it goes to your cell phone and you then call back and you don't leave a message, they have no idea it's you.
They may have had 10 or 20 other calls come in.
They're not going to go through and see what number corresponds to what.
So they think you didn't return their call.
They're going to think you don't care.
You don't want their business.
You don't want the job.
So get clear cut example.
What can work?
Calling your best friend or a family member can mean absolute career destruction if you do the same thing when it comes to how you use the phone in your business life.
So these are the things we're going to cover in this course of really learning this division.
I'm not telling you you have to use my system for every call in your personal.
You don't and you don't have to use the techniques I'm going to advocate in every single situation in your business life.
But when in doubt, I believe these are the basic principles that will help you the most and eliminate the chances that you've needlessly offended someone, confused a client, customer, prospect, boss

or coworker.

These are the principles that will protect you the most and help you advance your message, your agenda,
your business and your career.

3. Why Your Phone is the Best Secret Business App Ever :

We need to start off by recognizing the unique power of the phone and therefore the human voice, that's what the phone is good for, conveying the human voice.
And there are a lot of things that are distinct about that.
When you call someone, it shows a certain level of importance.
It's not just a quick email.
It's not just a quick text.
The phone is uniquely valuable for communicating things that are of extreme importance.
I want this job.
We want you to hire us.
We are firing you.
You've had a loved one who has died.
I mean, nobody wants to get an email or a text saying, will you marry me?
Or your child has died or your father has died.
It's considered too impersonal.
That's the beauty of the telephone.
It's much, much more personal.
It's a way of putting a spotlight on a message that you think is important.

It's a way for other people to communicate important things to you.
And that's why you're going to pay attention.
And I'm going to give you some techniques for really making sure it sticks with you.
But it's also why you're going to use the phone to communicate important things, to think about your own work day. How many emails do you get in an average day versus an average week?
Now, think about how many phone calls do you get? If you're like me or most people, you probably get a hundred emails, maybe a thousand emails for every phone call.
What stands out more now?
You can say you don't like phone calls or it's annoying, but if you're trying to do a serious, important part of business, get a new client. There's nothing like a phone call from a new client who says, yes, we want to work with you. Send over the contract.
Much more powerful than just an email, so always be thinking about not just how to use the phone, but when it is appropriate, there are times when a simple email or text will be fine, but for important information, serious stuff, things critical to your business, mission critical to your most important customers, clients, prospects, your bosses.
That's when if you can't meet in person, the phone is the next best thing

4. Planning Your Successful Phone Environment:

When you have an important phone call to make or you'll be receiving one, or if it's an important conference call, it's important to plan in advance, plan out your environment, to increase the odds that this is going to be an effective phone call.
For starters, you need to eliminate distractions, find a quiet place if you're talking to your most important client or you're pitching the most important prospect and you're in a noisy pit of an office.
And there are conference rooms that are quiet at the other end of the hallway or plan in advance and go down there before you make the call or before the call starts with someone calling you.
You want as quiet and environment as possible.
Now, in some offices or if it's a home office, you may have music on it that may relax you.
It may be soothing.
You may enjoy it.
That's all wonderful.
That's fantastic.
Guess what?
Nobody else cares about what you like for your workday.

If you're calling someone or someone's calling you, all they care about is their message, their ideas,
what's going on, their business, their life, their part of the job.
So eliminate distractions.
One reason you want to be quiet is if you have to talk over some loud noise that's annoying to listen to when you're talking over a loud noise.
For example, while driving a car, you're flattening your voice out.
You're making yourself monotone.
And it kind of feels from the other end like they're being yelled at.
Nobody likes someone yelling at them.
You're yelling because you need to hear yourself over just traffic noise, other distractions.
But to the person on the other end of the phone, all they know is you're yelling at them.
They can intellectually realize, well, us in a noisy street or noisy restaurant, but emotionally
it's still I'm yelling at them and it's not positive.
Let's talk about driving.
I understand we're all busy.
We all have busy lives were also important.
We all multitask.
Here's the thing.

If you're trying to have an important phone call and you really want someone to pay attention, don't be driving.
Now, I understand there are times someone's trying to sell you new aluminum siding and you're not sure you're even interested.
Yeah, you can tell that person to call you on your drive home.
And if you're only giving half way attention, it's not that big a deal.
It's not going to affect your business because you're not the one selling aluminum siding.
But when it comes to your business, what you're trying to sell, what you're trying to advance, what you're trying to promote with your vendors, customers, clients, prospects, investors, eliminate the distractions.
Because when you're driving, there are several things that happen.
There's the the hum of the car.
You can have the most fantastic electric Tesla.
There's going to be all the other sounds going by from other cars, highways, sounds honking.
So that's distracting to the person on the other end.
It's also distracting to you.
The other problem is your brain is seeing all this information.
You have to pay attention or you will die.

So you're giving much less attention to the person speaking to you.
So I realize it's tempting.
And if someone wants to start a conference call at five o'clock and you'd like to be on the road, then going home, the tendency is to get in your car a minute early and take the call in the car.
Don't do it when it's important.
Stay away from driving.
Stay away from noise.
Try not to be outside.
Hey, it's a beautiful sunny day one.
I take it outside.
The birds are singing.
Well, that's the problem.
The birds are singing.
There are too many distractions also within your office.
If you really, really are trying to focus on the person, get what they're talking about, be as responsive
as possible.
I'd go as far as turning off your computer, do what I just did.
I don't want to be distracted right now.
I literally turn my phone off.
I've got to use another phone, obviously, but if I'm using an office phone, turned my cell phone

off that way, if I get texts from friends about what to do Saturday night, I'm not looking down at that while talking on another phone.
It's not distracting.
The final thing, and some of you are going to hate me for this, but I'm going to give you my best advice.
You can do with it what you want.
If you really want to plan your environment the right way for making phone calls, don't ever use the speaker phone.
What are you talking about,me ?
We do all of our conference calls on speakerphone.
And how else are people going to hear?
Here's the problem with speaker phone.
The microphone is going to clip certain words.
Can't become kans.
There's confusion.
It's harder for people to understand you.
Number one.
Number two, you have to project your voice.
So you're kind of talking like this and you're projecting you sound like the school teacher.
It makes you kind of monotone flat.
You can sound angry.
It also tires you out.
Frankly, it takes a lot more energy to project if you're talking to a speaker on a conference table,

three feet, four feet, in some cases ten feet away. And it's.
Really cut down on the audio quality, frankly, nobody likes it when someone's talking on a speakerphone,
whether it's a big fancy speaker phone system in an office or just the speaker phone on your cell phone.
Again, there are exceptions.
You're trying to learn how to cook something and you're talking to a family member and you're in the
kitchen.
You need both hands.
I understand it.
I get it.
But when you're talking to people on business, they don't like it because the audio quality is worse.
It's harder to hear.
And the back of their mind, they're thinking, well, what's this person doing?
Are they still typing?
Are they reading the newspaper?
What do they need their hands for?
Many, many people interpret it, whether intellectually or just emotionally, as a sign of disrespect.

If you're on the speakerphone now, it raises the question, what do you do if there's five people in your office and everyone needs to hear, well, use a free conference call system, everyone can dial in and they can be on their phone in a different room, or you could even be on a phone, each person individually in the same room and hear it. Another option you can do is have it on speaker phone, if they're five of your colleagues, you have to listen to a boss or a client or a colleague.
But then when someone in your room is speaking, pick up the receiver and talk because everyone in the room can still hear.
But the quality of the audio for the person on the other end will be vastly superior.
Remember, most microphones in phones are built for maximum effectiveness.
If you're just inch and a half, maybe two inches away when you're further away, it really distorts the audio quality.
So plan your environment to increase your odds of effective communication before the phone call ever starts.
Whether you're sending a call, initiating one or receiving one.

5. Exciting New Update to this Course:

I've got some great news.
This course has been updated to be better than ever.
I've listened to so many of you tell me that while you like the course you don't like having to upload your speech your presentation your media interview to YouTube or some other video file sharing site and then cutting and pasting and putting that into the Udemy Q and A section so I've made it easier for you.
I've created a Facebook page that's just for my Udemy students.
It's just for you.
There's no sales going on.
There's no promotions.
It's just about your presentations your homework your questions.
So click below the link.
You can join and there's there's never any charge you'll automatically be accepted and that way.
If you want to you can post videos your homework right there with one click.
You can just hold your cell phone speak one click and it's right there on the Udemy homework page.
Now if for some reason you don't like facebook and believe me I know plenty of you and it's understandable.

You don't have to join Facebook.
I'm not coercing anyone.
You can continue to post your videos right here in the Q and A section of this course.
And there's absolutely no problem.
So either way the choice is yours.
But check out the link below.

6. Final Preparation for Your Successful Phone Meetings :

In addition to planning your environment in advance, before important phone calls, conference calls,
video conferences, it's important to plan everything else you might need because your client, your customer, your boss might be highly impatient.
They don't want to wait for you to go through a whole bunch of files or have to dig out something or get up and go to a file cabinet somewhere they want it right now.
So if you get any papers or documents you're going to be referring to during the phone call, the conference call, the Skype, have it all laid out on your desk.
Additionally, there's nothing like a good old fashioned pad and pen for taking notes during these calls.
And so you think, well, you know,., you old guy, look at that gray here.
Nobody uses pad and pen anymore.
I'll just type here's the problem with typing.
When people hear you typing, it's distracting.
And there's a part of you, part of them wondering what's this person typing?

Are they doing other business while they're talking to me?
Are they chatting with friends on Facebook or are they texting friends saying, you can't believe this idiot I'm talking to on the phone now?
It's just distracting.
This way, whether you have a headset or you're having to hold a phone, you've got one free hand, a pen, paper or pencil and take detailed notes.
Anything else you need, you're going to be showing slides through any form of a screen, sharing,
using zoom, Skype, have everything ready.
Respect the time of the people you're speaking to in business.
Doesn't matter if it's a co-worker, someone junior to you in another office, respect their time,
they will respect you.
And it's going to ultimately benefit as far as boosting your own reputation.
So plan get everything you need all together in advance of the call.

Section 2: Can You Hear Me Now?

1. Making Sure Your Phone Passes Your Friend Test :

You need to know exactly how your technology works.
I don't mean you literally have to open the back of the phone and figure out where every little like Trawick device connects but you do need to know the basics of what the most effective way is to use your phone.
My recommendation is call a friend.
In this case it could be a family member and ask them how do i sound and talk to it an inch away inch and a half two inches and on and give that get feedback from then on what actually sounds the best.
Because every phone can be a little bit different Your voice can be a little different.
You need to know what is the best way to use your phone as far as the distance from the microphone.
I can tell you right now the best way is not going to be two feet away with on speakerphone.
We've already talked about that.

I hope I've convinced you not to use speakerphone other than extreme situations where you absolutely have to have both hands free.
Couple of other things that are relatively modern problems with respect to phones especially for those of you who get new cell phones frequently as I do.
I had recently upgraded from one iPhone to the next.
Went to the same company that always sold me the phone I asked for a protective cover.
I got the protective governor and all of a sudden I'm noticing about a third of the people in speaking to are saying Ha why they couldn't hear me.
I kept on thinking it was because it was a new phone it didn't update.
Well no the company that sold me the phone and sold me the protective cover sold me the wrong version of a cover so it fit the right size but it was for the previous edition and the version they gave me covered one of the three microphones on the phone.
So it was destroying the audio quality took.
I'm ashamed to say it took me a couple of months to finally figure this out.
Very important when people tell you they're having trouble hearing you believe them do

whatever it takes to make sure you're coming across clear.
That's why I recommend testing.
That's why I recommend instantly trying to troubleshoot change things.
Make sure you don't have a cover that's obscuring one of the microphones so this is your first homework assignment.
Call a family friend a colleague could be someone in the next cubicle and ask them to tell you to give you honest feedback.
How does it sound to you sound muffled.
Is there anything unclear.
How is the volume really get feedback and if there's any problem at all let's figure out the solution right now so please don't just fast forward to the next video.
Take two seconds and call somebody.
And let's really get an assessment figure out how close you need to be to your phone.
And if there are any other problems with the microphone on your phone or if there's anything covering it.

2. Planning for Winning Conference Calls :

A great way to connect with team members who live and work in different cities different countries or clients spread out around the world is a conference call.
Here's some tips on how to make the most of your conference calls.
Now we've already covered.
Please don't use a speaker.
You can get tired of me saying that some of you will write the reviews he talked about not using speaker phones and yet people still do what I want to give you some other tips.
Realize a conference call is a business meeting and it deserves the same amount of planning and respect.
Of any other meeting too often people just oh let's just hop on a conference call and chew it out.
Yet they wouldn't ask people to drive across town and spend an hour meeting with no agenda.
So the first item on your checklist should be have an agenda know exactly what you want to come across with what you want to accomplish in any conference call.
Have some structure to it.
You'll come across much more effective.

Also realize if other people are talking to you and other cities they likely are going to be on a speaker phone because we can't control them.
If they are multiple people in the same room because of that it's going to be harder for you to hear Sometimes two people will talk at once.
They'll be cross-talk it'll be much harder.
So you've got to actively listen.
And on one hand you don't want to interrupt but you do want to interrupt quickly.
If you did not hear something if you didn't understand something so what I would recommend is relatively loudly saying excuse me I couldn't hear the last 20 seconds the audio dropped out or I couldn't hear because of the crosstalk.
Could you please repeat you do it like that.
It's not going to seem like rude interrupting and people actually appreciate the fact that you're trying to get the ideas conveyed and that you're trying to really understand them.
Now sometimes conference calls can get long and before you know it you're checking e-mail.
I've done it before myself checking Facebook don't do it.
Really focus.
Listen carefully.
Be Thinking visualizing what the person is talking about ask questions when there's a pause when

there's a time you don't want to be interrupting people on the phone whether it's a conference call or any other call.
But when they've said something you're thinking about it it is appropriate to summarize to paraphrase to make sure you're getting it.
Show them you're listening.
Show them that your understanding.
So you do periodically want to paraphrase summarize.
I'd also highly recommend taking detailed notes.
We talked about that in the previous lecture.
Take notes of any important phone call for business whether it's a conference call or anything else.
But here's something most people don't do but will make your conference calls much more effective.
Not only take your notes but then summarize the meeting from your perspective.
Summarize what you think everyone is supposed to do after that.
E-mail it to all the participants whether that's your official job whether anyone asked you to do that or not.
Everyone will be impressed.
Bosses colleagues customers clients you know who else this helps you.
We're all human beings.

We all forget things.
So the sheer act of writing it down and then typing it proofing it and then hitting send email will make all the key ideas from that conference call really sink in to you much more firmly.
This will help.
It's a little extra work but not much.
So please give it a try.

3. Look Prime Time Ready for Your Video Calls

These days it isn't just a phone.
It's a modern mini TV studio which means you can make phone calls using video.
There's FaceTime video there's Skype there's all sorts of different applications.
I'm not here to try to promote any one brand but I what I do want you to do is to think about how you can come across your most effective way possible when you are speaking to someone over the phone and you're using video.
There's so much more information conveyed when there is video people can see your face they can see whether you have bags under your eyes and if you haven't slept they can detect more about your emotional state.
So it is much more powerful.
I certainly prefer to speak to people through a live Skype video or a WebEx so I can connect with them where I can see them.
They can see me.
You might not prefer that it might not be used that often in your corporate culture or your organization your clients might not like it but if you do have the opportunity it's a very powerful tool.
I mentioned in the very first video of this whole course a couple of tips about the angle.

Again what makes so many people look bad and the reason they don't like video is they are afraid they're going to look awful stupid bad.
And if you hold your cell phone down here and it's shooting up your nose.
Horrible lighting it is going to look bad.
So I want to give you some techniques if you're going to use a cell phone or any mobile device for a video phone.
Ways of Looking a lot better.
For starters have the phone a couple of inches above your eyes now you can even put this on a bookshelf and you're standing and use walk a few feet away or if you're sitting down put it on a stack of books on your desk and aim it that way.
If all else fails just hold it like this.
Your arm might get tired but if it's only three minutes big deal.
Not suggesting you hold it like this for two hours.
You want to have more light in front of you than behind you.
I'm not suggesting you take all day long and come up with some elaborate lighting to read what I am suggesting is just look around your room if you're at your desk and there's a window behind you and it's a sunny day and you're holding the camera in front of you and the windows behind you.

It's going to look awful because the light is so bright coming from outdoors and your face is going to get all dark and that what I'd recommend just turn around have the light from the window hitting your face and you're recording from the opposite direction.
In general as long as you have more light on your face more light going this way instead of coming from behind you.
You'll look fine for any kind of telephone call Skype conference video conference with people.
Now we're back to the whole issue of audio because if you're holding your phone like this you're further away.
In most cases using ear buds the the microphone and headset that comes with a lot of phones you know older iPhones the mike is pretty good on that.
So you can just use that if you're going to be doing a lot of phone conferencing through video.
I would recommend you do what I do.
Again I'm not selling any brand today.
But I just use a simple lav Mike I can plug it on to my shirt jacket tie.
He's cost about 20 bucks from Amazon or any other electronic store.
So you plug it in and it's much more of a professional looking like professional sounding

Mike it plug it into a phone or whatever device you're using.
You really want to take it up a notch.
Now I do this more with my computer and iPad if I'm doing Skype but I actually use an earpiece.
That way I don't have to use the ear buds.
I'm using an iPhone.
This is just like what a professional broadcaster uses this Quick's on the back and it then goes right into my ear and you don't see it.
So it looks very professional Again this costs like 20 bucks on Amazon.
In simple earpiece and it's very effective you can do it for phones you can use it for iPads.
Any other computer.
So these are some of the basics you need to think about when it comes to using video phone calls.
The other big problem for most people is because they're self-conscious because they're nervous about who home we're going to look.
People freeze.
So if you have the best message Wolf well thank you sir.
We would love to do business with you.
We can do wonders for the marketing of Company X Y Z.
The message might be fine but would you want to do business with that person.

I don't think so because I looked scared stiff bored and boring.
So the biggest thing to remember when you're doing any kind of video telephone conferences you have to move.
You have to face.
You have to move your head.
You need to move your body and yes you even need to move your hands now your hands don't have to be visible to the person seeing you it may be much tighter shot than how you're seeing me now but the sheer fact that you're moving your hands means you're much more likely to be moving your body your head your face your eyebrows and you'll look like a real human being.
That's the key it's not about being beautiful or perfect or glamorous it's just about not looking like a scared stiff frozen robot.
That's what makes people scared afraid to do video.
Also really keep in mind the professionalism of what's behind you.
I have here is what I do when I'm doing a Skype video conference.
This is just a white PowerPoint slide on a TV screen.
You don't have to be that fancy.
But if you're on the road if you're doing escape video conference call and you're at a conference

and you're in a hotel room don't show and unmade bed behind who is just distracted looks unprofessional
and sometimes it's simply a matter of sitting on the bed and you're now showing the wall behind you.
No one can see the bed.
So really look at it see how other people see you on this video call because it is anything distracting.
I've seen people doing even videos here on you to me where behind them you see all kinds of liquor bottles and stuff.
Wow.
I wonder who this person likes to party all these distractions.
It's OK to have a professional office you know a lamp photos something like that.
Books are fine if you are in a library.
But just give it some thought.
Because human beings are easily distracted by any visual.
That's more interesting than looking at the face.
Give it a thought and I think what you'll find is video conference calls can be the most effective form of using the phone of all.
Once you get comfortable with it.

4. Placing Phone Calls Like the Consummate Professional :

Sanjit air can I speak to Sanjay.
That's not how you make a phone call.
Let's go over some basics now of when you are the one placing the call in a business setting.
For starters you need to sound pleasant.
No one wants to deal with someone angry grumpy rushed you might be having a bad day.
Nobody else cares.
So for starters you need to have a pleasant tone of what.
I don't mean Mr. Happy.
I haven't.
I don't mean anything fake cheesy but just a pleasant tone of voice.
And then you need to cover a few basics you need to state your name your organization while you're calling Hi my name is T.J. Jack with media training worldwide.
I'm calling Jande to follow up on the request for a proposal.
She asked me about last week regarding media training for the organization.
Is she available today.

Takes less than 20 seconds but I've covered the basics so I don't force the receptionist or some other administrative assistant or anyone else to go through all who is this and why is your car. Why are you calling.
I tell the person who you're with.
I made it all simple clear and I'm sending a signal.
I don't know who I'm talking to.
It could be an intern could be the CEO of a company.
I'm sending a signal that I'm a professional person that I'm respectful of the people on the other end of the phone and I like to be treated respectfully too.
So it sounds obvious it sounds basic but so few people do it.
They just say May I speak to so-and-so.
And it can be abrupt.
Now if you're calling and you're getting the person directly if all you do is say I was there and you're now talking to Jane she may be thinking well I don't recognize the number.
Who is this.
I could just say I'm busy.
Not a good time.
What are you do then.
If you've stated exactly who you are.
Maybe she was expecting your call didn't recognize the number.

Maybe she's happy to speak to someone from your organization about this issue even though she didn't know you were going to call so state exactly who you are why you're calling where you're from.
Let's start off with the basics.
I know it sounds obvious but it's often not done.

5. Always Knowing the Best Time to Call Clients and Prospects:

Hi Samantha This is T.J. Jack from media training worldwide.
Just following up on the email you sent earlier today asking for a proposal for presentation skills training.
Is this a good time to speak now for five minutes or so ask that person if this is a good time.
It's basic courtesy.
It might not be a good time but an hour from now might be ask.
It's great information shows you're respectful.

6. Now You will Never Wake Up a Client in the Middle of the Night:

When you're placing calls to clients prospects colleagues you need to be keenly aware of where they are in the world these days.
Even one person businesses are international You can have clients from any country any time zone in the world.
You don't want to call someone who goes to their cell phone and wakes them up at 3:00 in the morning even though it's 3:00 in the afternoon in your office so be aware if you're not sure then just go to Google and type in time in that city it'll tell you what time it is.
Obviously if some prospect calls leaves a message at 7:00 in the morning you don't know where they're calling from and you can't call back until 10:00.
No one's going to really fault you for that if they haven't alerted you as to where they are and you can't know for any other reason.
But if it's an existing client customer or prospect who tells you what city the area on or what country they're in you need to be aware of time zones.
You also need to be aware of their normal workweek.

Some of my clients in the Middle East they take off Friday and Saturday.
So if I call beyond 8:00 a.m. my time Thursday morning I've already hit their weekend.
So that's being rude to them.
So you need to be aware of that.
Other clients of mine even in the Middle East they have the same Saturday Sunday days off that we have here in the United States so don't assume it's something you can check in about two seconds on Wikipedia or Google but unless you're returning someone's call or calling in them for the first time hitting them on the weekend the better.
It's common sense on the one hand but if you're not used to dealing with people or people internationally it's not quite so common there.
Also we'll talk about this in the voicemail section. Some cultural differences with respect to voice mail.
There are some cultures where it's not just the young people it's people of any age just don't use voicemail.
You might find it annoying if you want to do work in their culture.
You'll have to deal with it you'll be better off sending an e-mail perhaps a text if you're calling and you do like I do and you do a lot of calls through Skype they're not going to see your number or figure it out so you get to figure out

some other way of letting them know you called or you reached
out to them because they won't see just by what pops up on the phone.

Section 3: Answering and Talking on the Phone like a Pro

1. Answering the Phone so Everybody Knows You are Ready for Business

Hello what do you want.
Obviously that's not how you answer the phone. Now there are different customs and different organizations.
What I think is the easiest for most people to do is just state their name and their organization. Someone calls me doesn't matter what time of day or night because I do use my personal five different phone numbers but I use my personal phone same as my office phone.

I have things forwarded to this unless I see it's from a friend or family member I'll just answer T.J. Jack media training worldwide.
This does a couple of things.
If I just say T.J. Jack and someone went to my Web site and they didn't see my name anywhere they just wanted to call and ask for proposal for media training or public speaking training.
If I just say T.J. Walker they're going to be confused.
They're going to wonder if they messed up.
Did they dial someone.
Wrong number did they bother someone by saying the organization.
It makes it easy for the person calling me.
So I do recommend you state your name and your organization.
Now some people highly recommend it.
How may I help you.
I don't do that.
It's not wrong.
If you want to I don't think it's essential.
The main thing is you want to sound pleasant.
You may be in a tense conversation with another colleague who may have just been fired.
You may be in a bad mood you may have a stomach ache.
Nobody else cares.
A prospect calling you doesn't care.

Colleague on the other side of the planet doesn't care.
They just want to talk to you now.
So be pleasant put a pleasant tone of voice on not super fake fancy's syrup.
But just a pleasant tone of voice and that lets someone OK.
I'm here for you.
That's all it takes when you're answering the phone.

2. No Such Thing As Answering The Phone Too Quickly These Days

Try to answer your phone as quickly as possible for most people if a phone call is going to their cell phone is right there it is pretty quick.
Some people still like to let the phone ring a whole bunch to somehow live like they're not too eager.
I don't think you can really answer a phone too quickly.
I don't think there's any negative vibes anyone's going to pick up on that.
I would suggest to you many cell phone plans have an option of how many rings it does before it goes to voice mail.
You don't want to make someone have to call and go through 10 12 rings before it hits their voicemail.
I would say after four rings if you can't pick it up let it go to voicemail.
That's less time that you're taking and the person trying to call you just makes it a lot easier.
So I would go and put your settings on that.
That's your homework assignment right now for your own fun.
See how many rings it takes before your voicemail sets.

3. Your Clients Will Never Think You are Screaming At Them Again

You are looking for the right kind of to see how annoying that is.
Here's the lesson.
Don't answer your phone if you're in a really noisy loud place.
Restaurants can be noisy.
Nightclubs could be noisy.
Far better to see a phone calls coming in.
Let it go to voicemail and then excuse yourself.
Go Outside.
Go to a quiet place.
Return the call five minutes later when it's quiet.
This way you can have a normal tone of voice you may have a client prospect calling you.
It's 3:00 in the afternoon for them but it's 9:00 p.m. for you you're it and you're off duty it's in a nightclub you're maybe having a good time having a glass of wine or two.
Don't show that side of your life your reality to the client or prospect at that moment.
Find a quiet place.
Wait till you're in a quiet place before you take the call.
You don't have to be shouting over music over background sound over the loud clatter of people if you're

to a noisy restaurant.

You want quiet when you're taking up most of the time.

Of course there could be exceptions where it's a client you're dealing with it's a very intense crisis situation you're going back and forth.

They have to get a hold of you instantly.

They've already talked a dozen times a day.

Yeah maybe you pick up instantly no matter where you are but otherwise you don't want to expose someone the first time talking to you and they're hearing you shouting it just does not make a good impression.

And also don't do that.

Be a good impressions if you're out.

Everybody deserves to relax in their own way.

But again had a couple of glasses of wine you at a nice loud fun place.

My recommendation just don't take the call.

Return the call tomorrow because you may think you're perfectly fine but if you had even a 5 percent slur in your speech that might make a permanent negative impression on a client customer or boss so make sure that you're not doing anything to expose yourself to criticism when you're taking calls after hours.

4. Knowing How to Sound Your Very Best - Every Time :

So you've initiated the call you're on the call.
How do you really come across.
Your best.
I gave you a few tips earlier and I want to reinforce those now for starters have a pleasant tone of voice.
No one wants to talk to a grumpy person.
Actually smile I know it sounds ridiculous and cheesy but actually smiling when you're talking on the phone will help.
You'll have a more pleasant tone.
Don't ask me why but somehow it works.
And if you even look at a mirror of yourself smiling you'll have a more pleasant tone standing.
You'll sound better you'll be able to breathe more fully.
You'll be able to gesture more.
You'll be able to resonate more.
So standing when you talk can actually help.
It's not essential but if you want to give yourself that final one percent extra It can be incredibly helpful when you're talking to a client customer or prospect.
Don't ever make them feel rushed.
I understand there may be times when you are rushed.

You can always apologize and in a non-Rush way say I apologize I have to leave now to catch a plane To be come see you for the meeting tomorrow.
Give a reason but otherwise than talk.
Don't rush also don't hop right in.
If you're talking to someone who is an existing client customer prospect or a boss and you already know them they can be seen as too rude too abrupt.
Have a little small talk ask how they're doing.
If it's a beautiful day where you your say how beautiful it is and ask how their day is now if you sense they're incredibly rushed or they're one of these people who always wants to get in and out and off the phone as quickly as possible.
You can dispense with the small talk but if you're not certain it's better to just spent 30 seconds 60 seconds on some pleasantries before hopping into the main purpose of your call.
It's got to be proportional.
If it's really just one quick question from a colleague in the Cleveland office Sure call up ask. Get it and get on.
But if it is a client a colleague someone else you report to and you know your answer talk to them for 10 minutes and you haven't talked in a few days.
Well 30 seconds 60 seconds.

It's proportionate.
It's only one tenth of the 10 minute phone call.
So try to have some small talk next.
How are you coming across.
And by that I mean that you're using a lot of fillers when you're uncomfortable.
This comes out when you're talking to clients prospects bosses.
When people are uncomfortable they're thinking what do I say.
And that's when the pillars the eyes the arms the ERS the nose the soles come out again and again and again my recommendation.
Record yourself and listen to it.
That's the homework assignment.
I've been very light on homework in this course so don't complain about the level of homework but this is not the sort of class where it really helps to just do multiple choice test and write a bunch of sentences you need to hear yourself speaking on the phone.
So the homework is I need you to call somebody have a conversation get a second cell phone or use your laptop or iPad to record just your part of the conversation.
Doesn't matter what the other person does.
And I need you to really count as your arms are as the nose and see are you coming across well beyond that.

Do you come across as somewhat interesting do you like the quality of your voice.
Do you sound fully conversational or do you sound like you're reading a script.
Nobody wants to sound like a script reader and nobody wants to listen to someone who sounds like they're reading a script.
How many of us get phone calls sometimes daily from some sort of tell a telemarketer and you can tell it takes less than a second.
To tell when someone's reading we all have built into us this instinctive ability to detect when someone's reading to us on the phone.
You don't want to sound like you're reading.
And by the way this applies to conference calls any kind of call I do not recommend you ever read on the phone to someone.
They can figure it out.
They can hear it.
Of course there are exceptions.
If you are doing a quarterly conference call and you are the CFO for a publicly traded company and you have to read legal boilerplate or your lawyers say you're going to prison.
By all means read it but realize at that moment you're not trying to really communicate you're just

checking a box for any other time when you're talking to someone and you've initiated the call specially you want to just talk.

You want to be conversational so I need you to record yourself having a conversation ideally work related with another colleague perhaps a friend a buddy appear an equal and just talk to them as if you were talking to a client.

Record it listen to it.

Figure out what you are like what you don't like.

That's the homework.

Do it right now.

5. The Right Way to Put Someone On Hold

Excuse me one moment I have a sick child at school and I'm just waiting on the call from the school may I put you on hold for one moment please.
That's how you should ask someone to be put on hold.
Don't just say hang on a minute oh I've got to take.
If you just say that what you're telling the client customer or boss is hey you're not really very important that anybody else even the telemarketers were bored than you.
It's very insulting to someone so my recommendation.
Don't put people on hold but if you have to ask and give them a reason.
If you do that then you come across as a very respectful polite.
It's much easier on their ego.
I have a better perception of you and they're not going to think you're rude.

6. Making Sure They Only Hear the Magic of Your Voice

The green room do you want to hear that.
I don't think so you're probably disgusted.
All right you're probably looking for the one star rating button right now.
Why do I do that I just want to point out to you when you're talking on the phone don't eat.
People can hear it.
It strikes them as rude.
Have I ever been guilty of that.
Yes but I try not to.
And I especially try not to when I'm talking to clients customers prospects and other decision makers involved in my own business.
Should I never do that.
I should never do that.
But we're all human beings.
So if I'm talking to my spouse or something and it's dinnertime and I'm traveling Yeah I might do that. Learn what's appropriate in business situations so you don't want to be sending any noises that are getting it can be confusing to the business person you're speaking to.
So eating can be a problem.
Drinking slurping you may think you're really quiet but the sipping sound maybe louder than you think.

So that's why I'm drinking my tea in between takes of shooting this course for you.
Not during the actual video other than to make a point there.
Typing is another one.
We've talked about that.
But when you're typing.
People can hear it in the back of their mind they're thinking oh she's handling email now or always talking to friends on Facebook typing don't make noise is the only thing someone should hear when they're
speaking to you on the phone is your voice.

Section 4: Voicemail Can Be Your Best Friend

1. Being a Professional to Every Generation, Regardless of Your Own Preferences

O K I've been holding off on this part of the Course now and I let it all hang out here and be totally candid with you.
There are in fact some generational issues when it comes to how the phone is used.
Different people have different generations use the phone differently.
I'm not here to advocate for anyone generally.
No one generation is better than the other.
It's just different.
But here's the thing.
If you are in business or running a business or an employee in a business with very few exceptions you're
going to try to appeal to people of all sorts of generations if in a bigger organization.
There may be people from 16 to 80 working in your organization if you have any kind of a business it could be manufacturing designer T-shirts.

Yet maybe most of your customers are 22 years old but you might have some that are 75 years old buying something for their grandchild.

Why needlessly alienate any customer or any prospect.

So I'm advocating that you essentially become multi-lingual fluent in generation speak when it comes to dealing with the telephone.

And you know different people are bad at different things.

Believe me I know people my age and older people who were pitching and screaming for 10 years refusing to use text.

Now if a client sends you a text question you should text back the answer.

So that's the fault of people my age and older who were grumpy stick in the muds not open to new things.

That was a problem that hurt them in business communication because let's face it these days you could be 62 years old and have a 28 year old boss.

And some corporations a 28 year old sends you a text at 3:00 p.m. They're going to want a text answer response.

If they're in another city or another location or maybe even down the hall they don't necessarily want

you to get up or walk all the way down there send them a big memo or call them.
They just want to text back.
So I want you to realize I'm not here is just a grumpy old man criticizing the young ones.
Old people can be very guilty of not adapting to new technology and new uses of the phone.
A lot of older people could say oh I don't want to do Skype video or zoom because I'm getting nervous about the fact that their hair is falling out and they've got wrinkles and jowls.
So every age group has certain blind spots.
So now that I've gotten that over the way let's talk about a big problem with a lot of young people in the workplace.
Again we talked to the very beginning of this course about the difference between private life social life and business life.
I know a lot of people I have young nieces and nephews where if I were to call them it would be considered just ridiculous absurd breach of etiquette almost obnoxious.
If I were to leave a voicemail message beyond the pale it will not be listened to and it will not be returned.
And we get along perfectly fine with fine relationship.
It's just that's not how people of a certain generation 18 19 or 20 like to communicate as

there are of course there exceptions and of course I'm making some generalizations here but I'm just trying to be helpful.
There are certain people of certain generations where you just don't answer the phone.
It's just not something you do.
You see someone called and that's kind of a reminder to send them attacks are you going to bump intoNthem later but you just don't answer the call.
Well if you do that in business you may be fine. Absolutely.
Every single person you deal with every investor every client every customer is your age group too.
How often does that really happen.
If you have clients customers colleagues investors other people Halling you pick up the phone.
Answer it if you're in a critically important meeting obviously don't. But if you're not doing something wildly important that takes 100 percent of your attention.
Answer the phone.
People hate it when they can never get a hold of a count representative or a marketing director
that's supposed to be in charge of their camp but they hate it.
They absolutely can't stand it.

If they're let's say over 40 or a certain age younger people may think no problem I'll send the text.

You have to realize everyone is different and if you're dealing with a customer client prospect they expect you to bend to them if they're the ones writing a check if they're the ones paying you in the form of being a customer or a client or a boss they expect you to bend to them.

They don't really care the way you want to communicate.

Sorry if that sounds cynical I just think that's human nature.

And we have to reflect that it's selfish for you to expect people to bend to your needs.

Again if it's your company and you want to run it the way you want to and you don't really care if you alienate people then by all means don't change.

But the first thing I really want to stress is you've got to answer the phone.

That leads us to the next topic.

Voicemail and there are generational issues here too.

But there are people who are guilty of all ages on bad voicemail etiquette that in the next lecture.

O K I've been holding off on this part of the Course now and I let it all hang out here and be totally

candid with you.
There are in fact some generational issues when it comes to how the phone is used.
Different people have different generations use the phone differently.
I'm not here to advocate for anyone generally.
No one generation is better than the other.
It's just different.
But here's the thing.
If you are in business or running a business or an employee in a business with very few exceptions you're going to try to appeal to people of all sorts of generations if in a bigger organization.
There may be people from 16 to 80 working in your organization if you have any kind of a business it
could be manufacturing designer T-shirts.
Yet maybe most of your customers are 22 years old but you might have some that are 75 years old buying something for their grandchild.
Why needlessly alienate any customer or any prospect.
So I'm advocating that you essentially become multi-lingual fluent in generation speak when it comes to dealing with the telephone.
And you know different people are bad at different things.

Believe me I know people my age and older people who were pitching and screaming for 10 years refusing to use text.
Now if a client sends you a text question you should text back the answer.
So that's the fault of people my age and older who were grumpy stick in the muds not open to new things.
That was a problem that hurt them in business communication because let's face it these days you could
be 62 years old and have a 28 year old boss.
And some corporations a 28 year old sends you a text at 3:00 p.m. They're going to want a text answer response.
If they're in another city or another location or maybe even down the hall they don't necessarily want you to get up or walk all the way down there send them a big memo or call them.
They just want to text back.
So I want you to realize I'm not here is just a grumpy old man criticizing the young ones.
Old people can be very guilty of not adapting to new technology and new uses of the phone.
A lot of older people could say oh I don't want to do Skype video or zoom because I'm getting nervous about the fact that their hair is falling out and they've got wrinkles and jowls.
So every age group has certain blind spots.

So now that I've gotten that over the way let's talk about a big problem with a lot of young people in the workplace.
Again we talked to the very beginning of this course about the difference between private life social life and business life.
I know a lot of people I have young nieces and nephews where if I were to call them it would be considered just ridiculous absurd breach of etiquette almost obnoxious.
If I were to leave a voicemail message beyond the pale it will not be listened to and it will not be returned.
And we get along perfectly fine with fine relationship.
It's just that's not how people of a certain generation 18 19 or 20 like to communicate as there are of course there exceptions and of course I'm making some generalizations here but I'm just trying to be helpful.
There are certain people of certain generations where you just don't answer the phone.
It's just not something you do.
You see someone called and that's kind of a reminder to send them attacks are you going to bump into them later but you just don't answer the call.
Well if you do that in business you may be fine. Absolutely.

Every single person you deal with every investor every client every customer is your age group too.
How often does that really happen.
If you have clients customers colleagues investors other people Halling you pick up the phone.
Answer it if you're in a critically important meeting obviously don't.
But if you're not doing something wildly important that takes 100 percent of your attention.
Answer the phone.
People hate it when they can never get a hold of a count representative or a marketing director that's supposed to be in charge of their camp but they hate it.
They absolutely can't stand it.
If they're let's say over 40 or a certain age younger people may think no problem I'll send the text.
You have to realize everyone is different and if you're dealing with a customer client prospect they expect you to bend to them if they're the ones writing a check if they're the ones paying you in the form of being a customer or a client or a boss they expect you to bend to them.
They don't really care the way you want to communicate.
Sorry if that sounds cynical I just think that's human nature.

And we have to reflect that it's selfish for you to expect people to bend to your needs.
Again if it's your company and you want to run it the way you want to and you don't really care if you alienate people then by all means don't change.
But the first thing I really want to stress is you've got to answer the phone.
That leads us to the next topic.
Voicemail and there are generational issues here too.
But there are people who are guilty of all ages on bad voicemail etiquette that in the next lecture.

2. Voicemail That Soothes, Not Angers

We have to talk about your voicemail.
Now again I tried my hardest not to sound like a grumpy old man in this course and I think I held off for the most part a little bit in the last lecture but I really got to let it all out here.
Many of you of all ages have a huge huge voicemail problem and it is destroying your credibility.
It is harming your career advancement.
It is shutting down your pipeline of new prospects customers.
Let's start with some of the basics.
OK you've got to have a voicemail on your cell phone.
If you don't have it it's pretty rare for people not to but every once in a while they encourage it and counter it.
You call and it just rings and rings and rings and then it goes dead.
I do understand there are some cultures in the world where it's just not the norm and there are even some religious cultural influences as to why that is.
So I don't mean to sound culturally insensitive and I appreciate that here on you to me.
We have students from 190 countries.

I personally have students from 183 countries at the moment of this recording.
And if you are in a culture where it's just not the norm to have voice mail.
I understand.
But I do want to give you a tip if you want to deal with people from other cultures.
You need to get voice mail if you want to deal with people who are used to having voice mail.
You're making it a lot harder for them to do work with you to feel comfortable with you because you're wasting their time.
If I have to call you 10 times before I get you and I can't leave a voicemail and I can't send you a text because I'm calling you through Skype and it doesn't work that way I could use what's up.
That's another way.
But you're making life difficult.
So if you want to increase the odds of communicating more effectively with customers clients prospects bosses.
My advice is number one you've got to have a voicemail.
Number two is going to sound obvious but it's not.
You have to actually check your voicemail again.
It's very common in my experience for people say under thirty eight just don't listen to voicemail.

You may have the function that record or takes the audio recording and converts it to tax and you read it that's fine.
I'm not saying you literally have to listen to it although in many cases the transcription isn't that great and it's not going to capture the sense of emergency or the emotional component.
If you listen so I do actually recommend you listen to the voice mail from Clintonite every telemarketer but an important client or a boss you've got to actually listen to it because if someone leaves you a voicemail and you don't listen to it.
How do you respond if you don't respond.
That's a slap in the face.
You're telling the customer client prospect you're not important.
I don't respect your time because I'm now going to make you tell me the thing a second time if I just call you and say hey what's up your client or customer or prospect may have left you a detail message about a specific problem they want you to solve.
And if you don't listen to it wasting their time.
It's a big world out there.
There are a lot of other vendors consultants people who do what you do.
Nobody has to work only with you.

So if you make life difficult for people they'll find someone who will make life easy.
So by all means you've got to listen to voice voicemails and respond to them in a timely manner.
Absolutely critical.
OK.
Next issue with voice.
You got to have an actual message that's personalized.
If I call you and all I get is the generic message.
You have reached 2 1 2 7 6 4 4 5.
First of all it takes a long time to hear all those numbers and the generic message number two.
I'm now wondering did I call the right person.
Now I got to look at the number I wrote down.
And if that's on the phone and it's the same phone it's difficult.
I've got to wonder is this person still in business.
If I'm calling someone who is when I think I was a highly successful Internet consultant or a real estate broker and I get just a generic number like that there's a part of me and I'm not the only one who thinks this is a part of me that thinks what they are afraid to put up their own name to had bill collectors chasing them.
Are they a criminal or a drug deal and what are they hiding from.

It just doesn't sound professional to get a random number.
Now I know a lot of people not picking on the young kids anymore.
I know people my age and older who are independent real estate brokers who are independent technology
consultants and they have that situation where it's just a random number.
And guess what.
They don't make much money.
They are some of the least successful people.
Nice people friendly people but they just don't care about their image they don't care about their reputation and they make it really hard for people to get a hold of them.
So you need a message and the message shouldn't be cutesy we don't need to hear your dog barking again.
If this were 10 or 20 years ago everyone had a clear differentiation between Here's my home number and my home voicemail or you know the recorder that played it back on the tape and here's the office phone it would be different.
But today things have blended together people's work phones personal phones cell phones.
Typically it's not one phone it all filters into one phone through call forwarding.

So you need something that isn't too cutesy that isn't just personal.
If people are calling you and it's a business call again Mine is very simple.
I'm T.J. Jack and I state the name of my company.
Leave a message I'll call you back.
It's that simple.
So here's my homework assignment for you I want you to have a friend call your voicemail and then tell you what they think.
Ask for any kind of feedback what are they like what do they not like.
So people try to be so curt and cute and say you've called T.J. Jack right click.
I mean that's cute for about two seconds the first time you hear it if the person is less than 23 years old.
After that if you're calling the person many times it becomes annoying.
Now they're really really big issue and this is a great way to get yourself taken off the list of someone seems a rising star is an advance for someone that we shouldn't do business with.
It's when your voice mail is full.
And I realize some of you are saying that T.J. Jack you're just so old nobody hears uses voice.
I'm here to tell you people he has voicemail maybe not your friends maybe not the people you date maybe not the people you socialize with.

Maybe not the people you graduated from college with in the last five years but people do like to leave a voicemail if they've taken all the time and trouble to find your name find your phone number dial letting it ring hearing your message.
They now want to leave a voicemail.
So if you've gone through all that after you've heard this big long message and then now you're ready to leave your message.
And you hear this recording the person's voice mail is fall and cannot accept new messages.
It is extraordinarily annoying.
I guarantee you your boss is not impressed.
Your new client is not impressed because now they've done all that work they have to hit and call and they've accomplished absolutely nothing but wasting their time.
Thanks to you.
So now they get to decide do they want to tax.
Well believe it or not not everybody makes every single phone call from a cell phone.
And this all really shocked some of you but I know plenty of people over 60 don't have smartphones.
They use flip phones.
They're not using text.
And if they need to get a hold of you and they're not near a computer because they can't send an

email or text through their phone they're just out of luck.
Do you want to write off all those people you might not want to party or socialize with people over 60 but they still have money and their money still works.
Any place else you want to spend it.
So I'm begging you have voicemail.
Check your voicemail.
Have a personalized voicemail that is professional that states your name your organization.
Make sure you take it all voice mails out after you've heard them delete it.
So there is space for new messages and then return because somebody calls you call back.
Now here's the message to old people somebody text you with a question text back to them.
Do this and you will be in great shape.

3. Best Voicemail Messages for Success

When you're leaving a voicemail it's critically important you do the following things.
For starters state your name your organization and the purpose of your call.
Speak a little bit slower than normal and articulate not in a big fake phony way but articulate a little more clearly than usual.
Give your phone number once and then give it a second time even a little bit slower.
If you do that you'll come across professional.
You'll make it easy for people to get back to you and the person you're calling you may think knows you by heart but they may know 10 other people with your name.
So unless you have a really distinctive name state your full name.
I mean your first or your last name.
We should still have to state your middle name but state your first and last name the organization.
Make it clear for them.
And if it's any kind of a call where you're not certain the person is going to want to call you back right away.
Perhaps a prospect dangles something interesting I don't mean cheesy or provide a comb if you want to get rich in the next 24 hour.

I'm not suggesting you do anything overtly manipulative and cheesy and filled with hokum but if you can give them a taste of what you're calling about and why it might benefit them to call you back you'll increase the odds that they call you back.
A lot of people they get a voicemail when they hang up.
They do not do that.
Get some of you are watching this and saying oh my gosh T.J. is a really old grumpy man.
I'm just trying to help you make more money and succeed in business.
I understand there are a lot of you under.
I'll say about the age of 39 where you never ever ever ever ever leave voicemail with friends and family.
Fine.
I'm not trying to change you.
But if you're calling a customer a client or prospect do not call and then leave a voicemail it's highly irritating to a lot of people they can't stand they're thinking about is this a telemarketer I'm wasting my time.
Or is this something or do I need to call back or what do I do it creates all this anxiety.
So if you get a call someone leave a message.
That's what the technology's there is literally nothing bad that can happen to you if you leave a

voicemail after you've called someone and it goes to voicemail.
It just makes life easier for everyone you don't have to do this constant game of phone tag or people trying to figure out why do we even start trying to contact each other we will all be there in your initial voicemail message so please I'm begging you call get the voicemail leave a message.
You may think you're calling an office phone but the office phone has been forwarded to someone's cell phone.
And if they just see a couple of missed messages it's annoying and they're going to wonder was this important is this crisis what's really happening.
Why keep them in doubt.
It's not necessary to do that at all so please leave the voicemail message.
It's simple.
The technology has been around a long time.
You cannot go wrong by doing that.

4. Take 1, Take 2, Take 3 Until You Get It Right

Final important voice mail tip many voicemail systems allow you to listen to it and redo it if you're Not happy.

Very few people ever follow up on that but do it now you don't want to spend 10 hours listening and redoing a simple message to your boss every day.

Eat up too much time.

But if it is someone who doesn't know you well there's not a strong preexisting relationship a client prospect maybe a brand new client and you're not sure maybe you had a bunch of ads hit the button that

allows you to listen to the message and then hit the buttons as if you're not happy hit this button and rerecord it.

Keep recording it and listening to it until you're comfortable with it.

This alone should take some of the stress off a lot of times people get nervous and that's when the UNs come out.

If you know in advance it is voicemail then it takes all the pressure off.

You can just read to it and tell you're happy with it.

Also I should have mentioned this earlier but it's critical so often these days we're just assuming it's going to be voice mail we can actually get a human being that we're often shocked to have an actual human being picked up.

That's the best of all possible worlds don't say oh I was hoping to get your voice mail.

Use the opportunity be ready for the opportunity to say exactly what you wanted to say don't make them hang up so you leave a voicemail that's awkward and frankly weird.

Don't do that.

OK here's your homework I need you to leave a voicemail on someone's phone and then call them up ask them to tell you what they think of your voicemail.

Get some feedback get a critique from someone else ideally in your place of business of what they think of the message you leave on your voicemail.

Please do that now.

Section 5: Final Telephone Success Tips

1. Never Be Busy for Clients and Bosses Again

Another small issue I don't encounter it very often and it does come up.
Call someone you get a busy signal who in the world still has a busy signal.
Now I do realize it costs sometimes a dollar or two more on an average cell phone plan to have it go right to voicemail spend the money.
Look if you're poor and it's your own personal phone for your personal time fine.
You have to have a phone.
But we're talking about here what you're doing for business and in this day and age.
If a client or a prospect a customer is calling you and they get a busy signal.
It sounds like you're out of business or you're a restaurant and somebody started in 1955 and they're just you know they're in the kitchen all the time in figuring anything out.
So get rid of the busy signal.
No one who calls you for business should ever have their time wasted by hearing and at and at.

2. Making Your Phone Disappear At the Perfect Moments

I haven't talked to you about one of the most powerful ways you could ever use your phone in business.
One of the most powerful ways you can cement your reputation with your boss poignance customers away.
This phone can enhance your credibility dramatically.
What's that secret.
The off switch.
What do I mean by that.
Well if you're in a meeting if you're in a face to face meeting with a client prospect customer an important colleague turn your phone off.
It's insanely annoying to people if they're talking to you and your family hang out with me I go take this and they don't know what it is.
For all they know you're just you know placing a bet with your bookie or something.
So my recommendation if you want to really send a signal to a client customer prospect that hey you are important your time is important your business is important.
My relationship to you and your organization is important.
Turn your phone.

I don't mean just on vibrate because if it's vibrating and lights are flashing but not ringing.
It's human nature.
You're going to look there you look and there's a part of you that's going to think Is there a chance this is more important.
It probably isn't.
But it's going to take you away.
It's going to send a signal to the person you're meeting with in person that they are not important.
Well guess what.
Nobody likes to be told they're not important and they might not be that important but you don't have to rub it in their face.
So know how to turn your phone off.
I would recommend you not put on the text so many people go to meetings these days and their cell phone is there and worst is when they start playing with it that's the absolute worst we've all been bored at meetings or conference.
It's one thing if you're at a convention there's a big keynote speaker and you're in the dark and you're on the 18th row and the speaker is not that interesting.
Well sure does check email you check the sports scores but if you're in a meeting and someone's talking

and you're on your phone it's frankly about the same as speeding.
You might not feel that way but I just realized other people do.
Do you like it when someone spits on you do you like it when someone tells you your time your ideas are completely worthless.
Now we're not here to talk about family relations and whether teenagers or other family members are on their cell phones at dinner time.
That's a different course.
I'm not going there.
I'm talking about just in business situations.
You really are not using the full potential the power of your cell phone unless you know exactly how to not only turn it off put it away.
In fact a really important meeting don't even take it in with you.
I understand life is messy and complicated.
There are exceptions but you need to plan for the exceptions and lay the groundwork.
So if you have a meeting with your boss.
But the most important client to your company is planning on calling you.
They said I'll call you tomorrow morning.
We've got to resolve this one issue it only take two seconds.
Then I would say to my boss I have this out.

Just because Mr. Smithers is calling it he has one quick question and I promised I'd be available to him.
Your boss is going to realize that Smithers is the person paying everyone's bills because it's the most important client your boss isn't going to be offended.
And have it out and let the person know.
You could also have something that any fair minded person considers a very important personal issue that should take precedent.
You could say Excuse me everyone.
I don't mean to be rude but close family members about to have a baby I'm expecting a call from the hospital at any time.
It's the only call I would take.
That's why my phone is out.
No one is going to be offended by that.
Or if you have a family member who is in the hospital said look I'm taking this out just because I might be getting a call from the hospital my father is in intensive care.
No one's going to be offended.
But if he does have the phone out and you don't say why and you're kind to her.
Yeah.
It's going to cut against your credibility.
You seem like someone who can't focus who's scattered who's just all over the place.

That is not a reputation that's going to enhance your career.
So my recommendation.
Any important meeting phone not just on vibrate but off preferably not visible.
Ideally it's not even in the same room with you.

3. Tit for Tat and Text for Text

You heard me picking on the answers now I'll pick on the oldsters if someone in business sends you a text with a simple question text them back you might not like texting.
You might not have a phone where the keyboard works for you get a new phone.
You want to be in business you need to have a phone you can type easily.
Have you retired all you want to do is talk to your grandkids.
Then use the flip phone from 1998.
That's fine.
But if you want to be serious about business and you want people to feel like you respect their time you've got to make it easy for them and you have to communicate the way they want to.
You may be a 49 year old account manager but your client is a 25 year old rep for a fashion company.
They send you a text for a question.
They don't want to get a phone call from you.
They want you to text the answer back.
So that's the use of the phone.
Use the phone the way your clients customers and prospects want you to use it communicate on their terms not your terms.

Section 6: You Can Be a Communication Skills Master

1. Quick Wins! The High Tech Way to Perfect Communication

Let's hop right in with a quick win.
There is a simple high tech solution that will dramatically boost your communication skills it's right in front of us and yet most of us never use it.
Here's the tool.
It's a cell phone.
But wait a minute.
Here's the trick.
If you have to communicate to a prospective boss about a job or a raise or you're trying to raise money from investors or you want to ask people for their vote for public office.
Here's the key.
Practice what it is you want to communicate on video.
Look at it.
Figure out what you like don't like.
Once you've got a version on video that you like.
Here's the thing that you're not going to have done before I promise you e-mail that video to a friend or colleague.

Ideally someone whose mindset is similar to the person you're trying to communicate with and then ask your friend call them text then ask your friend and say not what you think but what do you remember.
What messages do you remember.
If they remember the messages that were important to you.
Now you can relax.
You know you are communicating.
But if they're not getting it it's not their fault.
It's your full time to practice again come up with new messages a new style for communicating.
This one technology will really dramatically help you.
Most people are doing it but now you can.

2. Listening is Key (and why this isn't just a public speaking course)

You can be a great public speaker a fine all writer.
You can have tremendous charisma.
And guess what.
Still be a lousy communicator your communication skills can be poor because if one thing you're not listening this is a communication skills class.
Of course I'll cover a lot of fundamentals and advanced steps for public speaking and PowerPoint and presentation skills but so much of being a great communicator goes deeper than that.
And so many communication opportunities in life.
It's about listening.
The best communicators in the world are often the best listeners.
People who are really great at making the one on one sale the best sales pitch.
Sometimes the best politicians who get the votes.
It's because they are the best at listening.
Now later on in this course there's an entire section on listening skills to really help you build the listening skills.
But it's something I want you to think about right now.

It really applies to every single type of communication whether you're asking for a job or raise a budget to be approved.
You need to listen to what other people say.
I've been in countless pitch meetings with a client where PR firms were pitching them for their business and the ones that lost talk the most.
They had the most slides up.
They stood up and gave the most formal presentations the PR firms that won they ask questions.
They sat back and they listened and then they reacted quite often with another question.
So this applies to personal communications with friends and families colleagues bosses rather than just pushing out all your preordained ideas.
You can't really listen and great communication isn't something you just created working until midnight on a PowerPoint three weeks ago.
You're there present with the person or with the group you're communicating with you're constantly listening
recalibrating and adjusting and believe it or not.
This even applies to public speakers the best speakers in the world.
Pause they're looking for responses from their audience.
Sometimes it's just a nod.
It might be a look of confusion.

That's a type of listening.
If you are the speaker so we're going to be covering more on listening in this course.
Just realize so much communication has nothing to do with your lips moving or even your eye contact.
It's about really hearing the messages coming out from other people so we can respond to it.
Let them know we've heard them react to it and deal with their concerns ideally before we deal with our own concerns.

3. Become part of the top 1% of communicators right now!

A very famous movie star and director once said 90 percent of success is just showing up.
I want to apply that to communication now all of us communicate all the time with friends family but so many of us just stop communicating when we get outside of our comfort zone.
We're afraid to raise our hand sometimes in a classroom.
Sometimes if it's a boardroom and they're intimidating people around we're afraid to call that talk radio show.
We're afraid to post our comments on a newspaper comment section.
We're just afraid to communicate our ideas.
So I want to put something here right in front of this cause for you to chew on to really think about beyond all the skills of framing messages the right way.
Good I can't.
And we'll get into all those style and mechanics. So much of being a good communicator is just having the willingness to speak out even if you're not quite comfortable.
So I'd give you a tip.

Right now you can automatically zoom up to the top 1 percent of communicators for this class just by doing one thing.

I want you to communicate with me right now.

Go to the Q and A section of this course and you'll see it right at the top in the middle of the page.

Once you've clicked the dashboard and I want you to communicate you can do it in a text format and write and tell me what you want to get out of this course.

You can post a video you are l put it on YouTube post it here telling us a little bit about who you are what you want to get out of this court.

You can make an audio recording and upload it to a site and put the audio there.

You get to choose do what's comfortable for you.

But most of you sad to say are not going to communicate at all.

You're going to sit back passively like let's go to the next one.

Let's get through this cause.

Let's see if we can get to the bottom and get our certificate.

That's not really learning and it's not communication so asking you politely please communicate with me right now.

Let me know specifically what you want to do to get the most out of this course.

And even though this course is jam packed there's five hundred and fifty lectures in this course.
I can still delete want to add something new if more of you want that.
So communication has to be a two way street and to many instances in life whether it's in school whether
it's in a big corporation of big government.
We're used as just sort of sitting back passively receiving and we buy into this idea that we don't have the right or the credibility or the standing to communicate back.
Well that's just not true especially in the modern world of technology Internet social media.
It's never been easier to communicate in any format we want.
So please take a moment communicate something with me right here.
Even if you want to say D.J. you're born we get on with it.
That's OK.
Post your communication now.

4. Good news, the problems we think we have, aren't real

I've got some good news for all of you.
But first let's step back for a moment.
We all know that sometimes there's perception sometimes there's reality.
They might be exactly aligned but they may be skewed.
So for example my perception is that I will never become a star professional basketball player for the NBA the National Basketball Association my perception of that.
It's because I'm too slow.
I'm too old.
I don't shoot well enough.
And I'm that shy of six feet tall I'm not tall enough.
That is my perception of why I'm never going to be a star in the NBA.
Well it turns out that my perception of my flaws are exactly aligned with the reality.
Those are real flaws.
Those will keep me from ever becoming an NBA star.
Now enough of you by this course I become a billionaire maybe I can buy a team.
But that's different.
I'll never star as a basketball player because I don't have the talent the physical talents at all to

do what it takes so many times in life.
Our perception of our weaknesses are actually correct. I'm here to tell you when it comes to being a great communicator it's not correct.
I have worked with tens of thousands of people all over the globe in person.
Real life for more than 30 years.
So I hear time and time again people telling me Oh T.J. I'm not a good communicator.
I can't give good speeches or presentations for the following reasons.
I say too many times I say say too many of Oz ers I'm not good looking enough.
My accent is too Southern or too New York or it's too Indian or it's too different from what people are used to.
People have this sense of all these problems that are holding them back.
People have told me I know I'll never be a good communicator because I have Bell's Palsy and my face droops meanwhile no one else in the room even noticed it.
So in my experience as individuals we come up with all sorts of reasons as to why we'll never be a great communicator.
Oh I don't like my voice.
My voice is no good.
Here's what I found working with training and studying with the top communicators public

speakers presenters communicators TV hosts in the world not bad manners the things that make us great communicators.
It's not just we didn't say arm some of the most successful highly paid TV stars in the world site and constantly.
It's not necessarily about your looks.
Some of the most successful TV talk show host and reality star host have less hair than I do and are 30 years older than I am.
So it's not about being young and good looking so I'm here to tell you.
Relax.
The things you think are making you a bad communicator really aren't.
Those are not your problem.
I'm not suggesting you don't have some challenges to overcome but it's a different set of problems because trying to change your accent is really really hard and can take hundreds of hours sometimes thousands of hours.
Do you have an extra thousand hours changing the tone of your voice is extraordinarily difficult.
I'm here to tell you you don't have to do that.
I've seen masterful communicators who have high voices low voices screechy voices accents that are not considered desirable or attractive.
And these people are still great communicators so I want you before we really dive in deeply in this

course.
It just kind of set aside these preconceived notions you have about what's holding you back.
People think Oh I moved my hands too much.
I'm here to tell you I've worked with again more than 10000 people around the world face to face for 30 years.
I've never yet had an audience say about someone I'm training.
Oh their hands are moving too much.
Never ones happen.
I do hear people say oh gosh that person seems frozen and stiff and scared and nervous that I hear all the time.
So we're going to work on a number of issues to make you a better and better communicator.
Right now I just want you to relax.
If you've ever had one interesting conversation with one friend one family member one colleague about one subject and you feel good about it and the person you are speaking with felt good about it.
Guess what you already have all the skills you need to be a great communicator.
We'll try to build those out. Cv
Build your comfort level so you can do them in different situations.
But you don't have to learn a whole new accent or lowers your voice perpetually.

5. Communications skills are the most important skills for success

If you're this far into the course I probably don't have to sell you on the importance of communication skills if you didn't think you were important.
Why would you be here.
But I do think it's important to kind of step back for just a moment and put it in perspective.
When you look at the most successful people in any field whether it's politics business high tech finance even the sciences so much of what that person does all day long is they communicate they speak to investors to voters to employees to vendors to customers to clients.
They have to listen to the concerns of their investors of their customers of their clients of their patients.
The higher you go up in any carrier other than perhaps writing novels it's about how will you communicate in a non written way.
Now obviously writing is a form of communication.
That's the form of communication we're going to spend the least amount of time here in this course because it is something that's taught every day in school your entire life.

What's not taught is how to communicate by speaking verbal communication nonverbal communication.
Basically any form of communication other than writing or typing that's what isn't taught.
And yet ultimately that's what could decide whether your career ends here here here or down here.
So I want to congratulate you for devoting some time investing your energy your resources into improving
yourself as a communicator and I would just want to redouble those efforts for you because again you look at the successful people you admire.
They might not be silver tongued orators but they are good communicators.
Even someone you think of who is not your favorite politician doesn't give a great speech.
Even people like that you see them in a small room one on one or with 10 people.
They can have contributors eating out of the palm of their hands.
Now you may have no interest in running for office or ever asking for money but I guarantee you there are aspects of your life your personal life your professional life where you can benefit by being a
better communicator.

So that's why as we go through this whole course you're going to see some sections that will apply to you.
Some that don't pick and choose as you see fit.

6. Direct versus indirect communication

Be direct.
Get to the point.
Be concise tell it like it is.
Don't sugarcoat it.
Just get to the point quickly.
You've heard all these things your whole life right.
Well guess what.
It's not really true.
Now I'm not suggesting you lie beat around the bush but in every single opportunity when you're trying
to communicate with someone you have to look at a number of factors.
What's their interest in what you're talking about.
How much do they trust you like you even know you.
It's not true that the most direct communication is always going to be the most successful.
Now this is not a course on dating and romantic relationships.

But think back when you were in high school or middle school there was typically always one person typically a guy and he would just call or text the 10 most attractive people in the school and ask them out on a date.
Hey you want to go out with me.
What typically happened to that person.
They all said no.
It was just too direct.
He was clear on what he wanted but it just is too off putting.
So it's not in.
That's why you typically don't see 30 minute infomercials as 10 second commercials an infomercial is a form of communication.
It's one we like to make fun of and people like to think it doesn't have an impact on them.
But infomercials the 30 minute long commercials spend a lot of time making the case telling a story telling benefits bringing you in being interesting talking about your concerns before they ever ask you to pull out your credit card number and purchase something.
So I want you to go into this whole learning experience on communication skills with an open mind and realize it's not about being fast concise the most direct.
It's about putting forth your ideas in the warmest environment possible so they can be received in

the best possible way so that the person or persons you're speaking to really understand you and feel good about you and your messages so they're more likely to understand you.
Remember your messages perhaps do what you want go where you want to go or join you in some greater enterprise.
So I do want to just get you to kind of scratch off this idea that more direct is always better. Sometimes it is but sometimes it isn't.

7. Let's hop in with something new. Big Business Communications

We are ready to hop right into the meat of this course.
I wanted to start with something different some of you have read books on communication you may have taken other courses here on this platform.
I want to give you something that I'm quite confident you haven't learned before.
So this first whole section here is about a very particular type of communication.
It's how you as an individual consumer can communicate with corporations big corporations

little corporations small companies and sometimes even governments.

It's a different form of communication than just talking to an individual friend family member colleague person at the office.

So that's what this section coming up is all about. I think you're going to see some things here that you haven't thought of before that can really make your life a lot easier.

As a consumer we're all consumers these days and most of the time you may have great experiences whether

you like to shop on Amazon or eBay or Wal-Mart dot com.

But let's face it we've all had some bad experiences and we just feel powerless.

We just don't know how to communicate our frustration are we really going to file a lawsuit every time we think a company gave us a bad product.

No that's not realistic but it is realistic to communicate to those companies effectively. The next section will show you how.

Section 7: Customer Communication Skills For Your Consumer Life

1. Communicate your message to companies, big and small, to get them to take action

So in this section you're going to learn exactly how to communicate with corporations big corporations
especially when you feel that you need something redress.
Typically this is when something went wrong.
Although you could communicate to companies one you're incredibly happy with them.
I like to leave positive reviews on Yelp.
Other places google reviews for companies that I feel treat me well.
But let's face it that's easy.
The trickier issue is what happens when a company is sold you a product.
It's awful and they're making it difficult for you to get a refund or you've bought let's say a plane ticket.
Now the trip was canceled and they want to hold your money for three months before you get a refund.

There's all sorts of things where we fall through the cracks as customers and unfortunately there is the expression the squeaky wheel gets the grease.
We've heard that we know that this section I'm going to teach you quick easy simple ways of communicating your dissatisfaction so that you are the squeaky wheel.
There's going to be some new techniques here. I think you'll be surprised but you'll be very happy with how fast and simple it is to communicate your messages with corporations large and small.
So stay tuned.

2. One person can get a huge corporation to pay attention and make things

I'm here to tell you one individual can get a huge mega corporation.
Pay attention you can communicate with corporations and even get what you want.
If you go about it the right way.
Let me give you a case study.
This was earlier this year and I don't mean to sound like I'm picking on one company.
It could happen to any company at some point.
This happened to be with my local Sprint store where I bought a cell phone I bought many cell phones there.
So I purchased the phone.
I've had many phones from their purchased a case and I couldn't hear.
People could not hear me.
I thought there was a problem on the phone with an update.
I thought it would work itself.
It didn't.
I took it to Apple.
They said Oh the real problem is the case.
It turns out the Apple store didn't do anything wrong.

The sprint store had sold me the wrong case I had made the purchase based on their estimate.
I went back.
I was very polite nice.
I just said I'd like a refund.
They wouldn't give it to me.
And I was polite.
I called asked to speak to the manager went by let my business card.
Nothing.
I figured Do I really want to waste time writing letters.
It was only a Seventy dollar case but still it just stuck in my crawl because at this point I've bought many phones from this one sprint store here in Huntington New York outside of New York City.
I just thought it was a really poor way of treating a customer.
And I wanted to communicate that.
But how do you do that typically if you're in a store and you see a customer service.
You see people in line they're getting angry and they're yelling at some poor customer service representative and they're yelling sometimes they're cursing and it's just easy to discount that person.
Oh you're just being obnoxious.
Go away.

So that's not what I did.
I had a very specific strategy where I wanted to communicate my message in a thoughtful way a polite way.
And most important a way that would get results.
I want you to watch the video and I apologize it's a little bit long it's about five minutes long but this is a video I posted.
And then here's what I did.
I e-mailed it to the local store manager the Sprint CEO from their Web site the Head of Investor Relations and Public Relations from their website.
And it caused a massive snarl and a lot of phone calls.
I'll tell you about the result after this but first I want you to see the video and then I will tell you why I did it that way why I'm dressed that way.
Why.
The approach I chose worked so take a look at the next video then I'll explain it and then we'll figure out how can you do this in your own life.
Because I'm sure you've ever bought anything if you've ever used it services of a cable TV company or phone company at some point you feel like you were treated perfectly.
This is the solution for you.

3. This video ruined the weekend of Executives at a major Phone Company

this is a message for T.S. Stuart who is the manager of the Sprint store
at the Walt Whitman mall in Huntington New York T. Apologies for this impersonal form of communication but I've stopped by the store to meet with you.
I left my business card for you to call.
I have phone and left messages and no one's returned the call.
So I really don't have any other means of communicating.
So the reason I went to the store is I have what I thought was a small problem.
For years I've had an excellent relationship with your store.
I have purchased at this point half a dozen phones from you.
I have lived in the area for a little less than five years and have regular monthly contracts with Sprint for my phone my wife's phone.
I have purchased several phones I've always had fantastic service in fact I've recommended my friends and colleagues to use your store and would frankly like to continue doing that.
Here was my what I thought was a small problem.

A couple of months ago I upgraded to the latest iPhone 8 bought a LifeLock protective cover which I've bought in previous years from your staff and it's done an excellent job of protecting my phone because I tend to drop them.
So I bought a new LifeLock jacket.
I guess it was around 70 dollars or so and I noticed that my phone just wasn't working half the time people couldn't hear me.
People couldn't hear me on skype calls certain people I called couldn't hear.
I thought it was a problem with the iPhone.
I thought it was a problem with software I figured let me wait six weeks.
Get the latest updates from Apple.
Nothing works.
So I took the phone back to Apple thinking it was a problem with the phone.
They did a diagnostic.
They took it out of the case.
They said the problem is the case you're using a case for an iPhone 7.
It's concealing the microphones.
Sure enough we took the case off and the problem was solved.
I haven't had any problems sense since I was at the store and didn't want to drop the phone and break
it.

I asked them.
They didn't pitch me on anything I asked them if they had another case that would work.
So I purchased another case a few days later.
I then went back to your store told them about the problem and I said frankly look I've I've had great relations with your store but I really feel like your consultant gave me profoundly bad advice.
They sold me a lifelong case for a 7 phone and I had an 8 phone and covered up the mike and the first thing your staff told me was frankly that I was basically an idiot because the case would have worked if I had the case door open at the bottom how I would have known that was beyond me.
Certainly no one told me that.
Next they told me that I had not returned the case within the two week period.
Therefore there was nothing they could do.
Well there was no rational way I could know there was a problem with a case in two weeks it's not like the phone had broken there was no way for me to know that was the cause of the problem I again suggested the problem was your sales team sold me the wrong case they gave me bad advice.
All I'm asking for is a refund on the case and I in fact return the case next your staff told me that

I didn't really need a new case.
It was just Apple CID trying to sell me stuff.
Implying that one of your major partners Apple is behaving unethically.
I can tell you the apple store was so incredibly busy they would have been happy if I had just left.
I had to ask them for a case they weren't trying to sell me this at all and I just had to say this leaves a very bad feeling in my stomach.
I like to think I'm a rational person.
I'm a polite person.
I went in stating my case.
It was instantly dismissed.
I was essentially told I was a fool.
I was wrong Apple was unethical everyone was doing something wrong.
Other than you and your sales team that just doesn't ring true to me.
Now I've given your company many thousands of dollars at this point and I'm essentially told I'm a liar and you don't return messages when I stopped by the store.
My messages aren't returned.
When I leave a professional business card my messages aren't returned.
When I call and leave a message with staff so I am left with the conclusion that you just don't care about me and customers like me.
And that leaves me very sad.

If you have anything that you think I've missed I'm more than happy to hear from you.
Thank you very much.

4. Here's what really happened.

So I hope you had a chance to look at that video.
Is there anything special about other special effects.
No it's just like this.
It's just me talking and it took me five minutes I did it in one take.
But what's the result.
I posted it.
I think about a day went by.
All of a sudden I'm getting calls from the local manager.
The local manager's boss the head of public relations people and corporate from many states away.
I'm getting calls on the weekends.
And there was anything we can do for you Mr. Walker.
Can we give you more.
Can we give you more money.
I wasn't trying to shake them down.
I wasn't trying to extort them.
I just wanted to be treated fairly.

All I wanted was a refund for the 70 dollar Kit and I got it and had got it like that.
It wasn't some six week process.
If they want to give it to you right away they can.
Here's what I think really happened.
Their goal was oh my gosh this guy is causing a problem because he makes us look bad.
He's embarrassing us.
How can we shut him up because this could blow up on social media.
Now it didn't blow up on social media.
It was only seen by a couple of dozen people and most of those were probably executives at Sprint.
So let's back up a minute and walk through my thought process my thought process was.
Any reasonable person who heard my side of this story how I was sold a case it was the wrong case wrong piece of equipment I should get a refund would agree with me.
So I concluded that I really was on the right side of things.
A fair minded person would conclude I deserve a refund.
So the way I position the video is I'm thinking about not how can I really communicate with Sprint but how can I communicate with other consumers who watch the video thinking this case seems reasonable.
Wow Sprint's really awful.

Maybe we shouldn't do business with them.
So I don't want to make it sound so complicated like Oh it's three dimensional chess but it is a little more community complicated than simply communicating with the Sprint executives or the local manager because I am trying to communicate with the people they care about their customers and the fear of looking bad with their customers I think is what would motivate them to really take a second look at my particular case.
And in this case do the right thing and just give me the 70 dollar refund they probably would have given me five hundred dollars as a way of making up for the inconvenience.
I didn't ask for that and I'm not asking you to use this communications tactic to shakedown corporations or governments.
I want you to use it as a tool to just help level the playing field and to help you communicate more effectively.
So more specific things idea I put on a suit and tie. I do believe it's just harder for businesses to discount someone who looks like a respectable businessNman and I realize there can be sexist and racist element classes elements in that but I did what I thought
worked for me.

I happen to be an old man who owns suits and can look respectable so I put a suit and tie on.
So I wanted to look the part of someone who is a good customer.
Notice how I started off in a completely positive way.
I stressed I had nothing but good dealings with your company.
I'm a reasonable person I actually said that I tell them I've given their cause their their business thousands of dollars over the years which is true and they have that in their records.
And here's the thing that most people forget.
I tell them I want to be able to give them more business in the future.
Now assuming for a minute that I was wrong in this situation I don't think I was.
I just think a fair minded business would say here's a customer that's given us thousands of dollars of business in the past gives us business every year gives us business every month through the the Sprint service agreement is likely to give us thousands of dollars more even if we think he's wrong.
Let's give him the 70 bucks back just to placate him again.
I'm thinking I'm right but I'm hoping they're at least in the worst possible scenario looking at it from that perspective.

So I'm incredibly positive I'm not cursing.
I'm not angry.
I'm not impugning anyone's motives but I'm stating my case in a straightforward way.
Again all of this is because I want the PR people the managers the managers boss to be thinking oh my gosh she's making us look like idiots because he seems so reasonable.
Most people when they complain I'm sure you've seen people at the airport.
There's bad weather.
A plane is delayed and some passenger wants to scream at the hapless person behind the ticket counter and it's not that person's fault that it's raining or there's a windstorm.
You don't want to seem like that.
That's embarrassing for everyone involved.
And let's face it when we hear I know when I hear someone yelling at a clerk it makes me want the clerk to say no go away and you're not getting any refund.
I want to be the opposite of that.
And when you're trying to communicate with a big company I think you need to be the opposite of that low key low energy but calm with confidence.
State your case and almost with the tone of a bewilderment of Gosh I've been a good customer I want to be a good customer.

I don't understand why I'm being treated so poorly and what you'll find is people will jump through hoops because they don't want a reasonable person to look at that and think wow they really are treating
their customers poorly.
Other issues involve the distribution.
I didn't just email that to the manager.
I called up and asked for the manager's boss got that person's email and then I just went to the Sprint Web site looked in the press section and pulled an email address off of a press release and I think I found the section on the head of the press office and investor relations so I sent the email out for five different places.
Seeing everyone so everyone saw what everyone else saw.
And this actually went out on a late Friday or Saturday so I'm getting calls from people in corporate headquarters on a Sunday afternoon and they're happy to come into the office.
Now I feel bad a little bit ruining people's weekends but I feel a little bad.
I had to go to the store and call several times and get no attention.
So this is a tactic that I believe is extraordinarily effective when you are in the right.

You can't lie you can't make up things or at least I don't believe you should and state your case calmly make a nice video.
Anyone can write a nasty Yelp review which by the way I think I did that too.
Oh the other thing I posted this video on the Sprint Facebook page so you can just imagine all the different buttons it's pushing.
By the way I did all this in about a half an hour.
This is so much easier faster quicker than you know threatening a lawsuit and writing regular letters.
It's quick and it's easy so to sum up corporation big government agency anyone it has to deal with a large group of public has done you wrong.
State your case in a video is more powerful than just a text comment and then distribute that video on YouTube Twitter.
Yeah I put it on there.
Hey I put it to their attention on their Twitter account too.
You hit them on all their social media contacts and you email it to them.
You will get their attention.
You will communicate your message so I hope you have nothing but good dealings with all your corporate purchases.
But if not use this technique.
Very few people are doing it.

That's why it works part of any great communication doing something a little bit different from what everyone else does.
It cuts through the clutter.

5. Never Let Any Company Have the Final Word with you again

Let's sum up what we've learned in this section.
You do not have to let a big corporation a big government institute have the final word.
You can be communicating with them as easily as they can be communicating with you.
And in this modern era it's exciting because the playing field is level at least a little more level than it was because you can communicate not just with the corporation or the government entity but you can communicate with all of their customers or their constituents.
And that's what really makes it powerful.
What people have traditionally done is they've written a long angry letter that's 12 pages long and maybe handwritten that gets tossed in the trash or one quick nasty review on Yelp.
And people just figure and there's always some negatives it'll balance out or if you are a really big

corporation or a government entity people sue you all the time you have lawyers for that it's easy to discount that and just put it in a box.
But the fear of public humiliation does get people's attention again.
Don't use it irresponsibly.
Don't just say it because you want something for free.
You're going to ruin it for everyone because it won't have any impact.
If more and more people do this this works when the facts are on your side.
And when a fair minded person hears your story and the company's story they would side with you.
The other beauty of this is it's so easy to do and so quick you don't have to have a studio like I do a professional lighting just pull out your cell phone make a video.
Speak from the heart.
State your case briefly try to do it even shorter than I did in the five minutes and then posted your page YouTube send it to everyone in the company or the organization or the restaurant whoever it is you're trying to get attention to.
Believe me your message will be communicated.
Good luck.
But again please use responsibly.